THE SIGNAL-FIRST RECRUITER

SIGNAL > NOISE

How Top Recruiters Spot Real
Skill in a World of AI Resumes

*A No-Fluff System for Technical Screening,
IT Staffing, and Staffing Agency Placements That Stick*

SEJALKHIMANI

RAHULKALE

DISCLAIMER

This book reflects the authors' professional recruiting experience and is intended for educational purposes only. It does not constitute legal, employment, financial, or compliance advice.

Employment law, hiring regulations, and industry standards vary by jurisdiction and are subject to change. The frameworks, scripts, and screening methods presented here are practical tools, not legal guidance. Sample contractual language, offer letter clauses, and verification procedures are provided as illustrative examples only and must be reviewed by qualified legal counsel before use. Always consult qualified legal and compliance professionals for matters involving employment law, regulatory requirements, or organizational policy.

Compensation figures, salary ranges, and market data referenced in this book are illustrative and may not reflect current conditions. They should not be relied upon for hiring, offer, or budgeting decisions without independent verification.

The views expressed are the authors' own and do not represent any current or former affiliated organizations. The authors and publisher disclaim all liability for actions taken based on this content.

All company names, product names, and trademarks referenced are the property of their respective owners and are used here for identification purposes only.

First Edition: 2026

Contents

Preface

We didn't plan to become recruiters.

We were engineers building futures in technical roles when recruiting found us, not the other way around.

Over two decades, working first together and then on parallel paths, we learned this craft the hard way: through rejection, through long weeks, through candidates who ghosted and clients who dismissed us. We made every mistake this book warns against.

What emerged was a system: not a collection of tips, but a repeatable framework for screening accurately, communicating consistently, and placing reliably. We've taught this system to the recruiters we've managed and watched it produce results.

Five principles anchor everything that follows:

1. Evidence beats opinion.

2. Clarity beats volume.

3. Decisions beat loops.

4. Transparency beats churn.

5. People-first beats short-term tricks.

This book isn't theory. It's what works.

A note on scope: This book focuses on candidate assessment, client management, and placement skills that apply universally across the IT staffing industry. Topics requiring specialized legal or regulatory expertise are beyond our scope. For those areas, we recommend working with qualified legal and compliance professionals.

— Sejal Khimani & Rahul Kale

Why Good Agency Recruiters Get Stuck (And What Top Performers Know)

Every Monday morning, you look at your placement quota and do the math. You're three short. Again.

You know what's coming: more outreach, more screens, more submittals. You'll work 55 hours this week. And if you're honest? You'll probably still miss quota.

This isn't a motivation problem. It's a filtering problem.

If you've been recruiting for IT roles for at least two years, you've made placements. You understand the basics. You know how to write a Boolean string, screen a candidate, and submit to clients. You're not incompetent.

But something isn't working the way it should.

You're competing with five other agencies on the same req and consistently coming in second. Candidates look perfect on paper but fail technical interviews, so the client no longer takes your calls. You're hitting your activity metrics but missing your placement quota. Candidates are using you to get competing offers and then ghosting when it's time to close.

Or maybe it's simpler: you're working harder than ever with more outreach, more screens, and more submittals, yet the placements haven't increased. You're busy, but you're not winning.

The ceiling you've hit isn't your limit. It's the limit of what you've been taught.

The Industry Reality

Most IT recruiters struggle to find qualified candidates consistently. Some of that is genuine market scarcity. Roles requiring rare skill combinations or offering below-market compensation are legitimately hard to fill. But most of it? That's a filtering problem. The candidates exist. Recruiters just can't separate signal from noise fast enough to get to them first.

For agency recruiters, this challenge intensifies. You're not just competing with the market. You're competing with 3-7 other agencies working the same req. When many recruiters can't filter accurately, and you're up against multiple competitors who also can't filter accurately, it becomes a volume game. Whoever submits fastest wins.

Except that's not actually true anymore.

A decade ago, first to submit usually won. Clients reviewed resumes in the order received. Speed was the advantage.

That started shifting. Clients got burned by fast, inaccurate submittals from candidates who looked good on paper but couldn't execute in interviews.

Today, most clients would rather wait a few days for one accurate candidate than receive eight wrong submittals in 24 hours. First to submit now only wins if you're also accurate.

Most recruiters are optimizing for the wrong thing. They optimize for activity: candidates contacted, screens completed, submittals sent, speed of delivery.

Top performers optimize for **signal**.

They care how accurately they can separate real experience from AI-polished resumes in 90 seconds.   They care whether their

10-minute conversation can predict whether a candidate will pass the client's technical interview. They care whether the client says "this is exactly what I needed" instead of "pass, send more."

This difference between activity and signal separates recruiters who stay busy from recruiters who consistently place.

What Changed Over the Past Decade

Ten years ago, recruiting was keyword matching. If the job description said "Java" and the resume said "Java," you submitted.

It worked because candidates didn't know how to game the system. Resumes were mostly honest.

Then three things happened.

First, ATS systems got smarter. Candidates learned to keyword-stuff resumes. If a job description mentioned fifteen technologies, their resume mentioned all fifteen, regardless of

whether they had actually used them.

Second, interview prep got industrialized. Platforms like LeetCode and YouTube tutorials taught candidates exactly what to say in technical screens. Surface-level knowledge started sounding like depth.

Third, AI arrived.

Today, candidates feed their work history to ChatGPT and ask it to optimize for ATS systems. Result? Every resume looks polished. Every bullet point uses power verbs. Every skill is framed with impact metrics.

Candidates use AI to prepare for screens. They ask ChatGPT: "What questions will a recruiter ask about Kubernetes?" Then they memorize the answers. When you ask "Tell me about your experience with Kubernetes," they recite a perfectly structured response, even if they have never deployed a container in production.

AI makes everyone sound credible. Entry-level with exaggerated experience or senior with proven depth. AI polish makes them indistinguishable on paper and during surface-level screens.

The old playbook of keyword matching, surface questions, and resume review no longer works. Everyone looks qualified. Everyone sounds confident. You're submitting candidates who fail at technical interviews because you couldn't tell the difference.

Clients don't care that "the resume looked good." They care that you wasted their time. When you waste a client's time repeatedly, they stop calling you back. They give the req to another agency.

The Three Levels of Agency Recruiting

Every agency recruiter operates at one of three levels.

Level 1: The Speed Submitter

This recruiter's strategy is volume and velocity. If the req says "Python," they submit everyone with "Python" on their resume. If the client wants five years, they submit anyone with five years or reasonably close.

They're responding to what the industry taught them: "First to submit wins."

A decade ago, that was often true. Today, first to submit loses if your candidate fails the interview.

Level 1 recruiters have high activity metrics, low placement rates, and they burn out fast because they're constantly restarting searches. This is not a judgment, as many firms incentivize this behavior. The system in this book works within those constraints; it adds precision to the volume.

Level 2: The Careful Screener

This recruiter has learned that speed without accuracy is expensive. They run phone screens. They ask questions:

- "Tell me about your experience with AWS."
- "How long have you been working with React?"
- "What's your proficiency with Kubernetes?"

But today, candidates have prepped these answers using AI. They sound confident. They use the right terminology. They give structured responses.

And the recruiter can't tell if it's real or rehearsed.

Level 2 recruiters submit fewer candidates than Level 1, but their pass-through rate is only marginally better. They're more careful, but they're not yet accurate.

This is where most intermediate agency recruiters get stuck. They know speed alone doesn't work, but they don't know how to validate depth in 10 minutes.

Level 3: The Precision Filter

This recruiter asks questions about decisions, trade-offs, and failures.

Not: "Tell me about your AWS experience."

Instead: "Walk me through the most complex system you built on AWS. What was YOUR specific responsibility, not the team's, but yours alone?"

Not: "How proficient are you with Kubernetes?"

Instead: "Why did you choose Kubernetes over Docker Swarm for that deployment? What were the trade-offs?" Not: "What technologies did you use?"

Instead: "What broke during the migration, and how did you diagnose it?"

These are **judgment questions**. They expose whether someone made decisions or just executed them, whether they understand trade-offs or just followed tutorials, whether they've handled failure or only worked on successful projects.

AI can help candidates sound polished. It can't fabricate three years of architectural decision-making.

Level 3 recruiters submit fewer candidates than anyone else. But their candidates pass technical interviews at significantly higher rates. Clients start giving them exclusive reqs. They stop competing with seven agencies and start getting direct partnerships.

Why Most Agency Recruiters Never Reach Level 3

It's not because they lack intelligence or hustle. It's because the industry optimizes for the wrong metrics.

Most agencies measure calls per day, screens per week, submittals per req, time-to-submit.

All of that is Level 2 work. It makes you busy. It doesn't make you accurate.

You might think: "My manager tracks submittals per day and does not care about quality."

Here's the reality: short-term, you satisfy volume metrics. Long-term, you satisfy placement metrics. The best recruiters do both by filtering faster, not by lowering standards. When you can scan a resume in 90 seconds instead of 5 minutes, you can review MORE candidates while submitting FEWER.

What separates Level 2 from Level 3 is **pattern recognition in real time**. The ability to hear vague ownership language ("we implemented") and immediately know it means limited personal contribution. To spot AI-polished language and dig deeper with follow-ups the candidate hasn't prepped. To detect hesitation and understand what it signals.

You don't need to become a technical expert. You need to recognize PATTERNS of depth versus patterns of surface knowledge.

Pattern 1: Ownership language ("I designed" vs "we implemented")

Pattern 2: Trade-off reasoning ("I chose X over Y because...") Pattern 3: Failure literacy ("When it broke, I diagnosed...")

These patterns sound the same whether someone's discussing databases, infrastructure, or machine learning. You don't need to understand the technology to hear them.

When you operate at Level 3, your metrics shift.

Placement rates typically improve because you're submitting fewer wrong candidates. How much improvement depends on your starting accuracy. If you're currently at 20% interview-to-placement, meaningful improvement is realistic. If you're already at 50%, incremental gains matter more than dramatic jumps. The principle holds: accuracy compounds.

Clients start requesting you by name. Instead of spreading reqs across seven agencies, they give you exclusives.

Your compensation potential expands. Top agency recruiters consistently out-earn their peers, not because they work longer hours, but because they place more consistently. The market rewards accuracy, not activity. How much that translates to depends on your market, your firm's comp structure, and your baseline. But the pattern holds: precision increases earning potential.

The Long Game

Agency recruiting is demanding.

Industry estimates suggest turnover at staffing firms often exceeds 25% annually for permanent staff, which is significantly higher than most other industries. Many recruiters transition out within their first year. The ones who last five years are uncommon. The ones who build two-decade careers and progress to senior leadership are statistically rare.

That's not a complaint. It's context for what this book is actually about.

The methods here aren't designed for short-term wins that cost you nights and weekends. They're designed for sustainable performance: hitting quota without grinding 60-hour weeks, building client relationships that compound over years, developing pattern recognition that makes you faster and more accurate as you gain experience rather than just busier.

Volume-first recruiting works until it doesn't. At some point, the hours become unsustainable, the burnout sets in, and recruiters either leave the industry or plateau. The recruiters who build long careers, those still placing at high levels after ten or fifteen or twenty years, figured out something different. They learned to work with precision instead of just intensity.

That's what this book teaches.

What This Book Does

This book teaches you what other resources skip: **how to filter accurately when AI makes everyone look qualified.**

The next two chapters cover the mindset required to operate at Level 3. Chapters 4-7 teach the specific skills: detecting AI-polished resumes in 90 seconds, running 10-minute screens that expose rehearsed answers, extracting real requirements from clients in one 15-minute call, and closing candidates who are shopping you against competitors.

Chapter 8 shows you the system in action through real-world scenarios.

The Appendix gives you the scripts, templates, and checklists to implement everything immediately.

How quickly you move from Level 2 to Level 3 depends on your starting point and how consistently you apply it. Some recruiters see results in weeks. Others need months of deliberate practice. The timeline matters less than the direction.

Fair Warning About the Transition

When you first apply this system, your numbers may dip temporarily. You'll submit fewer candidates while you're learning to filter accurately. If your manager is measuring daily submittals, this creates short-term friction.

You have two options:

Option 1: Apply this gradually by using Level 3 filtering on 20% of your reqs while maintaining volume on the rest.

Option 2: Have a direct conversation with your manager about testing a quality-first approach on one req. Frame it as an experiment.

Both paths are valid. The goal is protecting your income while you build the skill. If your firm measures daily submittals strictly, test this approach on 1-2 reqs first while maintaining volume on others. Prove the results before going all-in.

Honest Reality Check

Not every agency supports this approach. Some firms optimize purely for volume. If you're in that environment, you have three choices:

Apply these tactics selectively on your highest-priority reqs. Build proof by showing your manager the data when quality-first filtering works. Or recognize the firm's model doesn't align with your long-term goals and plan accordingly.

This book can't change your firm's culture. It can only change your skill set.

When This Approach Doesn't Work

No system works universally. Here's when this approach may not apply:

High-volume, low-skill roles. If you're filling 50 warehouse positions by Friday, depth-filtering is overkill. Volume wins.

Clients who genuinely don't know what they want. Some hiring managers can't articulate requirements no matter how skilled your intake. You'll need to manage expectations, not perfect your filtering.

Markets in freefall. During mass layoffs or hiring freezes, the best filtering system won't create demand that doesn't exist.

Your first six months in recruiting. Build fundamentals first by learning the basics of sourcing, screening, and client communication before optimizing for precision.

This book assumes you have the basics and want to level up. If you're in your first six months, start with Chapters 1-2 and the Appendix, which will accelerate your fundamentals. Return to Chapters 4-7 when you're ready to optimize.

What Happens Next

Reading about change is easy. Implementing change is harder, not because people lack commitment, but because changing ingrained habits requires deliberate effort.

Some recruiters will apply one tactic from this book and see improvement. Others will install the entire system. Both are valid paths.

The knowledge is in the next chapters. How you apply it depends on your context: your market, your firm, your clients, your baseline skills.

Some recruiters will implement everything and see dramatic shifts. Others will use 30% of the tactics and get 80% of the value. Both outcomes are success.

The only failure is reading this, agreeing with it, and then doing nothing because "it wouldn't work in my situation." Test one tactic. Measure the result. Adjust.

The choice is yours.

If You Need Results This Week

If you picked up this book because you're behind on quota and need something that works this week, not next month, start here:

Go to Appendix Section 4 (Communication Templates). Find the scripts. Use them today.

Go to Appendix Section 7 (Troubleshooting Guide). Find your problem. Apply the fix.

These two sections are your life raft. The rest of the book explains why they work. But if you need to swim first and learn later, start there. Come back to Chapter 2 when you have breathing room.

The Signal-First Mindset

The shift from Level 2 to Level 3 isn't about learning new tools. It's about changing how you think about your work.

Most recruiters measure themselves by activity: calls made, screens completed, submittals sent.

Top performers measure themselves differently: Did the candidate pass the technical interview? Did the client say "this is exactly what I needed"? Did I solve the problem or just create motion?

This chapter covers the mindset shifts that separate recruiters who stay busy from recruiters who consistently place.

Shift 1: From Activity Metrics to Result Metrics

Your manager tracks activity. Calls per day. Screens per week. Submittals per req. Those metrics exist for a reason: they are easy to measure and create accountability

But activity metrics don't predict placements.

You can hit every activity target and still miss quota. You can make 60 calls, run 15 screens, submit 8 candidates, and still place zero if you're filtering incorrectly.

Recruiters who consistently place don't ignore activity metrics. They satisfy them faster by optimizing for results.

Activity thinking: "I need to submit 3 candidates today to hit my weekly target."

Result thinking: "I need 3 candidates who will pass technical interviews. If the pipeline doesn't have them, submitting mediocre candidates wastes time because I will be restarting this search in two weeks."

Activity thinking: "I ran 12 screens this week. Good productivity."

Result thinking: "I ran 12 screens and identified 2 strong candidates. The other 10 looked good on paper but didn't demonstrate depth. I filtered correctly."

Activity thinking: "Client rejected my submittal, so I need to send more options."

Result thinking: "Client rejected my submittal. My filtering missed something. What pattern am I not seeing?"

When you optimize for results, you hit activity metrics faster because you're not chasing dead ends. You make fewer calls because your targeting is sharper. You run fewer screens because your resume filter is accurate. You submit fewer candidates because you've eliminated the ones who would fail.

Volume goes down. Placements go up. You hit quota in 40 hours instead of 60.

Shift 2: Self-Education as Competitive Advantage

Every week, the tech landscape shifts. New frameworks launch. Companies migrate platforms. AI tools change how engineers work. Most recruiters respond by hoping the hiring manager explains it during intake, Googling terms during screens, or relying on candidates to educate them.

That keeps you at Level 2 forever.

Top performers educate themselves continuously, in small doses, without waiting for a role to force it.

This doesn't mean becoming technical. It means understanding the landscape well enough to ask intelligent questions and recognize credible answers.

You see three reqs mentioning "LangChain" and "vector databases." You've never heard those terms. Instead of winging it, you spend 15 minutes researching: LangChain is a framework for building LLM applications. Vector databases store embeddings for AI/ML systems. You don't need to understand the architecture; you need to know what problems they solve.

Next day, you screen a candidate claiming LangChain experience. Instead of "How long have you used LangChain?" you ask "What problem were you solving with LangChain, and why did you choose it over building a custom solution?"

You can now tell if the answer is substantive or rehearsed, not because you are technical, but because you understand context.

Client gives feedback: "Candidate didn't have enough production LLM experience." Now you know what to screen for next time.

This cycle of encountering an unfamiliar term, researching context, applying it during a screen, and refining based on feedback builds technical credibility without learning to code.

Time investment: 10-15 minutes daily. Compound effect: massive.

After six months of this discipline, you're no longer guessing during screens. You recognize patterns. You know which questions expose depth.

Most recruiters avoid this because it feels like extra work. It is, at first. After the habit forms, it takes 10-15 minutes daily and pays back hours in faster screens. The alternative of staying at Level 2 and competing purely on volume is far more exhausting long-term.

Self-education isn't about perfection. It's about staying slightly ahead of where you were last week. That edge compounds.

Shift 3: Speed of Trust Through Speed of Communication

Trust isn't built through charisma. It's built through **consistent, fast communication**, even when there is no news.

Most recruiters communicate when they have something to report: "Found a candidate." "Client wants to interview." "Offer extended." When there's nothing to report, they go silent. That silence destroys trust faster than bad news ever could.

In the absence of communication, people assume the worst. Candidates assume you've forgotten them. Clients assume you're not working the req. Both start looking at alternatives.

The 24-Hour Trust Protocol With

candidates:

- Initial contact: Response within 24 hours (same day when possible)
- Screen completed: Feedback within 24 hours
- Interview scheduled: Confirmation same day, reminder day before
- Interview completed: Debrief within 24 hours
- Offer extended: Daily check-in until acceptance • Offer accepted: Weekly check-in until start date **With clients:**
- Req received: Clarifying questions within 24 hours
- Candidate submitted: Follow-up within 24 hours if no response
- Search progress: Weekly update even if "no new submittals yet" Notice what's NOT required: good news.

You don't wait until you have a perfect candidate. You update on cadence: "Screened 8 candidates this week, 2 worth submitting, sending profiles by EOD tomorrow."

This consistent presence builds trust through **reliability**. People trust what's predictable.

What to Say When There's No News

Most recruiters struggle here: "I haven't found anyone. What do I tell the client?"

Say exactly that, with context. ×

Weak: [Silence for 2 weeks]

× **Weak:** "Still searching, market is tough."

✓ **Strong:** "Update: Screened 12 candidates this week. 3 had cloud experience but lacked security background. 4 were junior despite listing 5+ years. 5 were shopping multiple agencies and not serious about your comp range. Adjusting sourcing to target security-first roles. Expect 2-3 qualified profiles end of next week."

This shows you're working the req, understand the requirement, and sets expectations transparently.

Clients don't expect perfection. They expect **visibility**.

Speed of Communication as Competitive Advantage

When you're competing with other agencies, speed of communication is your primary differentiator. If you're the only one providing weekly updates with substance, you become the preferred partner, even if you are not first to submit.

A candidate applies on Monday. You respond Thursday. By Thursday, two other recruiters have already screened them. They're in final rounds elsewhere. Your req just became their backup option.

A client asks for an update. You wait until you have "real news" to share. Three weeks of silence. They assume you're not working the req. They give it to another agency.

Speed isn't about being frantic. It's about being predictable.

In both scenarios, you lost not because you weren't qualified or didn't have good candidates. You lost because someone else communicated faster.

The Response Time Hierarchy

Not every message deserves the same urgency.

Immediate response (within 1 hour): - Candidate no-shows for interview - Client changes interview time - Offer acceptance or decline - Urgent problem that affects the search

Same-day response (within 4 hours): - New candidate applications - Screening requests - Interview feedback - Req intake

Next-day response (within 24 hours): - Candidate questions about benefits - Client questions about candidate background - Follow-ups on submitted candidates

Weekly response: - Progress updates - Market insights - Pipeline status

This hierarchy lets you stay responsive without constant interruptions.

Communication Templates

See Appendix Section 4 for complete communication templates. Use them to maintain speed without sacrificing quality.

How to Batch Communication Without Losing Speed

Block scheduling:

- 9-10am: Respond to overnight candidate applications
- 12-12:30pm: Mid-day check-ins (interview confirmations, follow-ups)

- 4-5pm: End-of-day candidate feedback and client updates

This gives you 2.5 hours of dedicated communication time spread throughout the day. You're responsive without being reactive.

Use automation strategically:

- Auto-acknowledge applications
- Auto-remind candidates before interviews
- Auto-prompt yourself for follow-ups (CRM reminders)

But never automate personalized communication (screen feedback, offer discussions, client updates). That's where relationship lives.

The Long-Term Compounding Effect

Fast communication doesn't just win you one placement. It builds a reputation.

Candidates who had good experiences refer others. Clients who get consistent updates give you more reqs.

After two years of consistent fast communication, you're no longer fighting for attention. People seek you out because they know what to expect.

That's the competitive advantage: not being the smartest or most experienced recruiter, but being the most reliable.

Note: This effect is strongest with direct clients. In VMS or high-turnover client environments, focus your trust-building on candidates because they follow you across jobs and become referral sources regardless of which client you're serving.

Time investment: 2-3 minutes per update.

Trust dividend: exponential.

The Candidate Experience Principle

Treat every candidate like a future referral source or a future hiring manager, because statistically, some of them will be. The senior engineer you reject today may be the VP of Engineering making vendor decisions in five years. The candidate who ghosts you may warn three friends. The one you treated with respect and transparency, even when delivering bad news, remembers that.

This isn't ethics for ethics' sake. It's compounding returns on professional reputation. Every interaction either deposits or withdraws from your credibility account. The recruiters who build sustainable careers treat candidates as long-term relationships, not transactions, including the ones who don't get the job.

The Three Shifts in Practice

Signal-first thinking across a typical week:

Monday: New req. Don't start sourcing immediately. Spend 30 minutes clarifying must-haves, research unfamiliar tech, identify filtering criteria.

Tuesday: Screen 5 candidates. Three can't explain trade-offs or decisions. Filter them out. Submit only 2 who demonstrated depth. Update client: "Screened 5, submitting 2 strong matches, filtered 3 who lacked ownership."

Wednesday: Client rejects one submittal, interviews the other. Ask for specific feedback on rejection. Adjust filtering. Update candidate before interview.

Thursday: No interview results yet. Check in with candidate and client proactively.

Friday: Interview went well, client wants second round. Prep candidate for what second round typically focuses on. Document what worked in this candidate's background for future pattern recognition.

Same activities. Different thinking. The result-focused recruiter optimizes every interaction for accuracy and trust.

Over time, this compounds:

- Better filtering = fewer wasted submittals
- Faster communication = clients prioritize your candidates
- Continuous learning = faster pattern recognition
- Result focus = hit quota with less activity

The Shift Doesn't Happen Overnight

If you've operated in activity mode for years, result-focused thinking feels unnatural. You'll question whether submitting 2 candidates instead of 5 is right. You'll feel awkward updating clients when there's no progress.

That discomfort is normal. And temporary.

After two weeks, the new patterns feel natural. After a month, your placement rate improves. After three months, clients request you specifically.

The shift happens when you stop measuring your day by how busy you were and start measuring by how accurately you filtered, how transparently you communicated, and whether you're slightly more knowledgeable than yesterday.

That's the signal-first mindset. Not louder. Not busier. **More precise.**

Clarifying the Timeline: When Metrics Dip, and When They Accelerate

In Chapter 1, we warned that your numbers might dip temporarily when you first apply this system. Here, we've said that optimizing for results helps you hit activity metrics faster. Both statements are true; they just describe different phases.

Phase 1: Calibration (Weeks 1-4)

You're learning new filters. Your 90-second resume scan isn't 90 seconds yet; it is 3 minutes. Your 10-minute screen takes 15. You're submitting fewer candidates because you're filtering more carefully, but you haven't yet built the speed that comes with pattern recognition.

During this phase, activity metrics may dip. That's the learning curve, not the system.

Phase 2: Acceleration (Week 5+)

Your filters are calibrated. Pattern recognition is automatic. You scan in 90 seconds because you've seen the red flags hundreds of times. You screen in 10 minutes because you know exactly what questions reveal depth.

Now you're hitting activity metrics faster because you're not chasing dead ends. Fewer wrong turns means more time for right ones.

The initial dip is an investment. The sustained acceleration is the return. Plan for both.

→ **Ready to apply this?** See Appendix Section 5 (4-Week Implementation Guide) for the daily practice schedule.

Flow With Trends, Stand on Principles

Every few years, the recruiting industry panics about the "next big thing" that's supposed to make recruiters obsolete.

First it was job boards. Then LinkedIn. Then ATS automation. Now it's AI.

None of them eliminated recruiting. They just changed what recruiters needed to be good at.

The recruiters who survived every shift had one thing in common: they adopted the tools without abandoning the judgment.

They understood which parts of their work could be automated and which parts required human discernment. They flowed with the trends but stood firm on the principles that actually matter.

This chapter covers how to operate in today's market, where AI handles commodity work and humans handle complexity work.

What AI Should Handle

AI isn't a threat. It's infrastructure. Like email or ATS systems, it's a tool that should save you time on repetitive tasks so you can focus on high-value work.

But most recruiters either ignore AI completely or use it wrong.

Here's what AI handles well:

Resume parsing and initial screening. AI can scan 300 resumes and identify which ones match basic keyword criteria in seconds.

This doesn't replace your judgment; it narrows the field so you're reviewing 30 candidates instead of 300.

Boolean search expansion. Instead of spending 20 minutes crafting the perfect search string, you describe what you're looking for and AI suggests multiple variations. You pick the best one, refine it, and run it. Time saved: 15 minutes per search.

Outreach message drafting. AI can draft personalized InMail based on a candidate's profile. You edit for voice and accuracy, then send. The drafting takes 30 seconds instead of 5 minutes. Multiply that across 50 outreach messages and you've saved 4 hours.

Interview scheduling. AI-powered scheduling tools handle the back-and-forth of finding available time slots. You approve the final time, but the coordination happens automatically.

Salary benchmarking. Instead of manually researching comp data, AI pulls real-time market rates for specific roles and locations. You review for accuracy, but the research is instant.

Follow-up reminders. AI tracks when you need to follow up with candidates or clients and prompts you. You still write the message, but you're not relying on memory or spreadsheets.

Notice the pattern: AI handles the **setup, research, and coordination**. You handle the **decision, refinement, and communication**.

This is the right division of labor. AI saves time on tasks that don't require judgment. You invest that saved time in tasks that do.

What AI Cannot Handle

Here's where most recruiters get it wrong: they assume that because AI can draft an outreach message, it should also send it. Because AI can parse resumes, it should also decide who to screen.

That's where quality collapses.

AI can assist with preparation. It cannot replace judgment.

Here's what must stay human:

Lie detection during screens. AI can suggest questions. It cannot detect when a candidate is reciting memorized answers versus explaining real decisions. That requires listening for hesitation, vague ownership language, and whether their story holds together under follow-up questions.

Pattern recognition across candidates. AI can identify keyword matches. It cannot recognize that three candidates with identical skills have completely different levels of depth. You learn this by screening hundreds of people and noticing what strong answers sound like versus weak ones.

Client relationship navigation. AI can draft a "no news" update email. It cannot read the client's frustration level, decide whether to push back on unrealistic requirements, or know when to escalate versus when to wait. That requires understanding context, politics, and relationship history.

Candidate motivation assessment. AI can track whether a candidate responds to your messages. It cannot tell whether they're genuinely interested or just shopping for competing offers. You assess this by listening to tone, asking about their decision timeline, and noticing what they ask about versus what they avoid.

The "unfillable req" call. AI can show you market data proving a role is mispriced. It cannot have the difficult conversation with a client where you recommend pausing the search. That requires credibility, timing, and the ability to frame bad news constructively.

Trust-building through communication. AI can remind you to follow up. It cannot build trust. Trust comes from consistent, transparent communication, which requires knowing when to update even when there's no progress, what level of detail to share, and how to frame challenges without making excuses.

The common thread: these are all **judgment calls under ambiguity**. AI operates on patterns from past data. Human judgment operates on context, nuance, and real-time adaptation.

When you automate judgment, you become a commodity. When you use AI to automate preparation and invest the saved time in better judgment, you become more valuable.

How Top Agency Recruiters Use AI

The best recruiters treat AI like an assistant who handles research and drafts, not a replacement who makes decisions.

Monday morning: New req comes in for a Senior Data Engineer. Instead of manually searching LinkedIn for an hour, you describe the requirements to an AI search tool. AI generates 3 Boolean variations. You pick the strongest one, run it, and get 40 candidates in 10 minutes. You just saved 50 minutes, which you will use for deeper screening later.

Tuesday: You need to send outreach to 20 candidates. AI drafts personalized messages based on their profiles. You review each one, edit for voice, adjust the hook, and send. Time per message: 90 seconds instead of 5 minutes. You've saved 70 minutes.

Wednesday: Screening a candidate who claims Kubernetes experience. Before the call, you ask AI: "What are the most common trade-offs when choosing Kubernetes over Docker Swarm?" AI gives you a summary. Now during the screen, when you ask the candidate about their decision-making process, you know enough context to tell if their answer is substantive or surface-level. AI prepped you. You did the validation.

Thursday: Client asks for salary benchmark on a Machine Learning Engineer role in Austin. AI pulls real-time comp data: $140K-$180K base for 5-7 years experience. You review it, cross-check with your recent placements in that market, and confirm it's accurate. You send it to the client with your interpretation: "Market data shows $140K-$180K. Your budget of $120K puts you in the bottom 10th percentile, which explains why we're getting pass responses. Recommend adjusting to $145K minimum to be competitive." AI gave you the data. You provided the strategic guidance.

Friday: Following up with 15 candidates you screened this week. AI reminds you who needs follow-up and drafts status updates. You personalize each one based on what you remember from the conversations, add any new information, and send. Time saved: 30 minutes.

Notice the workflow: **AI handles volume, you handle value.**

The Principle That Never Changes

Trends come and go. The recruiters who thrive through every shift understand one principle: **tools change, but the core work doesn't.**

The core work has always been: **accurately assess whether someone can do the job, and build enough trust that they choose you when they have options.**

No tool can do that for you. Tools can save you time. They can give you better data. They can automate coordination.

But they cannot replace:

- The ability to hear when someone is exaggerating their experience
- The judgment to know when a client's requirements are unrealistic
- The communication skill to update transparently when there's no progress
- The credibility to push back on a bad req without damaging the relationship
- The pattern recognition that comes from screening 500 candidates and noticing what depth sounds like

These skills aren't going away. If anything, they're becoming more valuable as AI makes everything else easier.

Where Most Recruiters Go Wrong With AI

There are two mistakes:

Mistake 1: Ignoring AI completely. These recruiters still manually parse every resume, craft every Boolean string from scratch, and spend hours on tasks AI could do in minutes. They're working twice as hard for the same results. In a competitive market, that's unsustainable.

Mistake 2: Over-trusting AI. These recruiters let AI screen candidates without human review, send automated messages without editing, and rely on AI-generated insights without validating them. They've traded time for accuracy. Their submittal volume goes up, but their placement rate crashes.

The right approach: **use AI for speed, not judgment.**

Let AI handle the first pass. You do the second pass. Let AI draft the message. You refine it. Let AI pull the data. You interpret it.

This gives you the efficiency of automation with the accuracy of human judgment.

The Competitive Edge

AI helps you compete on speed without sacrificing quality. When you're up against 5 other agencies on the same req, the one who can screen faster and submit more accurately wins. AI gives you speed. Your judgment gives you accuracy. Combined, that's the competitive edge.

The Speed-Accuracy Resolution

You might notice an apparent tension in this book: we've said clients prefer accuracy over speed, but we've also said you lose opportunities when someone else communicates faster. Here's how these reconcile:

Speed of communication is non-negotiable. Respond fast, always: to candidates, to clients, to every touchpoint.

Speed of submittal is contextual. Submit only when you have signal, not noise.

The recruiter who responds to a client inquiry within 4 hours but submits candidates in 48 hours with one strong profile beats the recruiter who submits in 4 hours with five weak ones. Fast response. Accurate submittal. Both.

The Current Reality

The recruiting industry is divided. Some recruiters use AI effectively by automating commodity work and investing saved time in better judgment. Others ignore it or over-rely on it.

The gap between recruiters who use AI strategically and everyone else is widening.

If you're not using AI to save time on research, drafting, and coordination, you're competing with one hand tied behind your back.

If you're using AI to make decisions for you, you're outsourcing the only thing that makes you valuable.

The recruiters who win understand this balance: **automate preparation, never automate judgment.**

Flow with the trend. Stand on the principle.

→ **Ready to apply this?** The next chapter gives you the filtering system that lets you use AI for speed without sacrificing judgment.

Filtering Candidates Fast
(From 300 Resumes to 3 Screens)

You open a new req. Within 24 hours, you have 300 applications. Your manager expects submittals by end of week. You have 4 days to screen and submit. That's 75 candidates per day to review if you did nothing else.

But you have other reqs. Other screens. Other clients.

This is where most recruiters break. They either skim all 300 and submit based on keywords (speed, no accuracy), screen everyone who looks decent (accuracy attempt, impossible timeline), or freeze from overwhelm and miss the deadline entirely.

None of these work.

Top recruiters use a different approach: **layered filtering**. Two fast passes that eliminate 90% of candidates before you invest serious time.

This chapter shows you both layers: the 90-second resume scan and the 10-minute screen. Master these, and you turn 300 applications into 3 strong submittals without working weekends.

Layer 1: The 90-Second Resume Scan

You cannot deep-read 300 resumes. You don't have time. Neither do the other agencies competing with you.

The 90-second scan isn't about finding the perfect candidate. It's about eliminating the clearly unsuitable ones so you can focus on the maybe-rights.

Red Flag 1: Buzzword Overload Without Context

What it looks like: "Led microservices migration using Kubernetes, Docker, AWS, Terraform, Jenkins, GitLab, Python, Node.js, React, PostgreSQL, Redis, Kafka..."

Why it's a problem: No single person touches 12 technologies deeply in one role. This is either a team effort where they participated minimally, or AI-generated keyword stuffing.

The 90-second check: Do the bullet points explain WHAT they built, or just list tools? If there's no outcome, no scale, no responsibility clarity, then it is surface.

Red Flag 2: Vague Ownership Language

What it looks like: - "Involved in the implementation of..." "Participated in the design of..." - "Contributed to the migration of..." - "Supported the team in..."

Why it's a problem: These phrases hide limited responsibility. "Involved in" could mean they attended meetings.

The 90-second check: Count the "I" statements versus "we/team" statements. If everything is collective, their individual contribution is unclear.

Red Flag 3: Title Inflation

What it looks like: "Senior Software Architect" with 3 years total experience, or "Lead Engineer" whose responsibilities read like junior-level execution.

Why it's a problem: Startups hand out inflated titles. Consulting firms promote fast. Title alone means nothing without scope validation.

The 90-second check: Does the responsibility match the title? A "Senior Architect" should be making system-level decisions, not just writing code.

Red Flag 4: AI-Polished Language

What it looks like: Every bullet starts with a power verb. Every line includes metrics. The language is perfectly structured but lacks specificity.

"Spearheaded cross-functional initiative to optimize cloud infrastructure, resulting in 40% cost reduction and improved system reliability."

Why it's a problem: This reads like ChatGPT output. The metrics feel manufactured. There's no story, just marketing copy.

The 90-second check: Is there anything specific enough that you couldn't write the same sentence about a different company? If not, it's generic polish.

Red Flag 5: Job-Hopping Without Pattern

What it looks like: 5 jobs in 4 years, each at a different company, with no explanation.

Why it's a problem: Could be layoffs (legitimate). Could be performance issues (red flag). Could be someone who gets bored and leaves (flight risk)

The 90-second check: Is there a pattern? Consistent layoffs during market downturns are different from quitting every 8 months during growth periods.

Green Flags: What Good Looks Like

Specific outcomes: "Reduced API latency from 800ms to 120ms by implementing caching layer using Redis."

Clear ownership: "I designed the authentication system. My team implemented it based on my architecture decisions."

Reasonable scope: 3-5 technologies per role, with clear depth in 1-2.

Logical progression: Each role builds on the previous. Responsibilities expand over time.

Context-rich descriptions: You can visualize what they actually did, not just what tools they used.

The 90-Second Scan in Practice

SCAN 1: Job titles (5 seconds): Does progression make sense?

SCAN 2: Current role bullets (30 seconds): Specific outcomes? Clear ownership?

SCAN 3: Skills section (10 seconds): Reasonable number? Depth or breadth?

SCAN 4: Red flags check (30 seconds): Buzzwords? Vague language? Title inflation?

SCAN 5: Decision (15 seconds): Screen / Maybe / Pass

After scanning 20-30 resumes, you should have 5-8 candidates worth screening. The rest are filtered out.

This isn't about perfection. It's about efficiency. You'll occasionally pass on a good candidate. But you'll save hours of wasted screens on candidates who were never going to work.

Layer 2: The 10-Minute Screen

You've narrowed 300 resumes to 8 candidates. Now you need to identify which 2-3 are worth submitting.

The 10-minute screen isn't a full interview. It's a depth check. You're testing whether their resume claims hold up under basic questioning.

The 5 Core Questions

Question 1: Ownership "Walk me through the most complex project you've worked on. What was YOUR specific responsibility, not the team's, but yours alone?"

What you're listening for: "I designed X." "I decided Y." "I was responsible for Z."

Red flag: "We implemented..." "The team built..." (deflecting to collective)

The Ownership Test: If you don't hear "I designed," "I decided," or "I was responsible for" in the first 4 minutes, the candidate is a contributor, not an owner. That's not disqualifying. However, if the role requires someone who makes decisions, not someone who executes them, you have your answer. Stop the screen. Move on. This single filter saves 15 minutes per wrong candidate.

What answers actually sound like:

Strong: "I designed the authentication flow. My team implemented it, but the architecture decisions (session management, token refresh logic, the trade-off between stateless and stateful) were mine."

Weak: "We built a microservices platform. I was involved in the authentication piece."

Rehearsed/AI-polished: "I spearheaded a cross-functional initiative to modernize our authentication infrastructure, resulting in a 40% improvement in security posture." (Sounds impressive. Says nothing specific. No trade-offs. No decisions. No ownership evidence.)

Question 2: Trade-offs "Why did you choose [Technology A] over [Technology B]?"

What you're listening for: Acknowledgment of downsides. "We chose Kubernetes because of scaling needs, but it added operational complexity we had to manage."

Red flag: "It's the industry standard." (No actual decision-making)

Strong: "We chose Kafka over RabbitMQ because we needed higher throughput for our event streams. The trade-off was operational complexity: Kafka requires more infrastructure expertise. We accepted that because our team already had Kafka experience from a previous project."

Weak: "We used Kafka because that's what   the architect recommended."

Rehearsed: "We leveraged industry-leading messaging infrastructure to enable real-time data processing at scale." (Marketing language. No actual trade-off reasoning.)

Question 3: Failure Handling "Tell me about something that broke in production. What was the impact and how did you diagnose it?"

What you're listening for: Takes ownership. Explains diagnostic process. Learned something.

Red flag: Blames others. Can't remember details. Claims nothing ever broke.

Strong: "Our payment service went down during Black Friday. I traced it to a connection pool exhaustion issue: we were opening connections faster than we were closing them under load. I implemented connection pooling with proper timeouts. We lost about 20 minutes of transactions, maybe $15K in revenue. After the fix, we load-tested to 3x our peak traffic."

Weak: "We had some production issues but the DevOps team handled most of that."

Rehearsed: "I proactively identified and resolved critical production incidents, ensuring 99.9% uptime." (Resume language. No specific incident. No diagnosis process. No learning.)

Question 4: Evolution "If you rebuilt that system today, what would you do differently?"

What you're listening for: Specific improvements based on lessons learned.

Red flag: "I'd do the same thing." (No growth mindset)

Strong: "I'd split the monolith earlier. We waited until performance forced us, which meant we were refactoring under pressure. If I did it again, I'd identify the bounded contexts in month three, not month twelve."

Weak: "It worked, so I wouldn't change anything."

Rehearsed: "I would implement more robust CI/CD pipelines and enhance our testing coverage." (Generic improvement. Not connected to actual lessons from the project.)

Question 5: Decision-Making "What's the hardest technical decision you've made in the last year? Walk me through how you approached it."

What you're listening for: Process for evaluating options. Considered multiple approaches. Can explain why they chose what they chose.

Red flag: Can't think of one. Describes a decision someone else made.

Strong: "Whether to migrate to Kubernetes or stay on ECS. I built a comparison matrix covering operational overhead, team expertise, cost projection, migration risk. Presented both options to leadership with my recommendation. We went with ECS because the team didn't have Kubernetes experience and the timeline didn't allow for learning curve."

Weak: "The architect decided we should use Kubernetes, so we did."

Rehearsed: "I led the strategic evaluation of cloud-native orchestration platforms, ultimately driving adoption of best-in-class container infrastructure." (Buzzwords. No actual decision process. No trade-offs considered.)

Pattern Recognition: Deep vs Surface

Dimension	Deep Experience	Surface Experience
Ownership	"I designed," "I built"	"We implemented," "Involved in"
Specificity	Numbers, constraints, timelines	Abstract descriptions
Trade-offs	Acknowledges downsides	Decision seems obvious
Failure	Takes ownership, learned	Blames others
Evolution	"I'd do X differently because..."	"I'd do the same"
Business context	Connects tech to business	Stays purely technical

The 10-Minute Screen in Practice

Minutes 0-2: Brief intro, confirm interest in the role.

Minutes 2-4: Question 1 (Ownership): Listen for "I" vs "we."

Minutes 4-6: Question 2 (Trade-offs): Listen for reasoning, not just answers.

Minutes 6-8: Question 3 or 4 (Failure or Evolution): Listen for growth mindset.

Minutes 8-10: Question 5 (Decision-Making): Listen for process.

By minute 10, you should know: - Did they make decisions or just execute them? - Can they explain trade-offs or just list technologies? - Do they take ownership of failures or deflect?

If the answers are surface-level, don't submit. You'll be wasting the client's time and damaging your credibility.

→ **Templates for this:** See Appendix Section 1 (Resume Filtering Checklist) and Section 2 (Screening Questions).

Getting Real Requirements (The 15-Minute Intake)

Most recruiters fail before they start sourcing.

The intake call is where the damage happens. You ask generic questions. The hiring manager gives vague answers. You leave with a job description but no actual understanding of what success looks like.

Then you source, screen, and submit, only to have every candidate get rejected for reasons that were never discussed in intake.

The 15-minute intake solves this. Five questions that extract real requirements, not job description bullet points.

Why Traditional Intake Fails

Most recruiters ask: - "What skills are you looking for?" - "How many years of experience?" - "What's the salary range?" These questions are unproductive.

"What skills?" gets you a laundry list of technologies that no single human possesses.

"How many years?" gets you an arbitrary number that doesn't correlate with ability.

"What's the salary?" gets you a range that may or may not be competitive.

You leave the call knowing what's on the job description, which you already had.

The problem: you don't know what success looks like. You don't know what failure looks like. You can't filter because you don't know what you're filtering FOR.

The 5 Questions That Change Everything

Question 1: "What will this person accomplish in their first 30 days?"

This forces the hiring manager to think beyond keywords. They have to describe actual work.

"Set up the CI/CD pipeline for the new microservice." "Audit the existing authentication system and document vulnerabilities." "Ship the first iteration of the recommendation engine."

Now you know what the job actually IS, not what technologies it involves.

Question 2: "What's the hardest part of this role that most candidates underestimate?"

This surfaces the hidden requirements. Every role has them.

"The codebase is 10 years old with minimal documentation. They need to navigate ambiguity." "The team is fully remote across 4 time zones. Communication discipline is critical." "The product manager changes priorities weekly. They need to stay calm under shifting goals."

Now you know what to screen for beyond technical skills.

Question 3: "Describe someone who failed in this role or a similar one. What went wrong?"

This reveals the landmines. What behaviors or backgrounds lead to failure?

"Last hire was technically strong but couldn't work without detailed specs." "Previous person got overwhelmed by the pace. They needed too much ramp-up time." "Someone who was great at building but terrible at maintaining." Now

you know what to filter OUT.

Question 4: "If you had to cut the requirements in half, which ones would survive?"

This separates must-haves from nice-to-haves. Most job descriptions list 15 requirements. Maybe 3 actually matter.

"Kubernetes experience is nice-to-have. Someone who understands distributed systems is non-negotiable." "We can teach the framework. We can't teach debugging instinct." "Security background matters more than cloud experience." Now you know where to focus.

Question 5: "What would make you reject a candidate after the first interview, even if their resume was perfect?"

This reveals the hidden filters that never make it to job descriptions. "If they can't explain trade-offs. I want people who think, not just execute." "If they talk about technologies but not problems. I don't need a resume reciter." "If they seem like they'd need constant direction. This role requires independence."

Now you can predict rejection before it happens.

The 15-Minute Intake in Practice

Note for agency recruiters: This assumes direct hiring manager access. If you're working through an account manager or VMS system, request that these questions be gathered on your behalf, or ask for a brief 10-minute call with the hiring manager. The information is worth the ask.

Minutes 0-2: Quick context (team size, reporting structure, timeline)

Minutes 2-5: Question 1 (First 30 days): What is the actual work?

Minutes 5-7: Question 2 (Hardest part): What is the hidden requirement?

Minutes 7-10: Question 3 (Failure pattern): What should I filter out?

Minutes 10-12: Questions 4 (Cut in half): What actually matters?

Minutes 12-15: Question 5 (First interview rejection): What is the hidden filter?

After 15 minutes, you should be able to answer: - What will this person actually DO? - What separates someone who will succeed from someone who will fail? - What's the hidden filter that's not on the job description?

If you can't answer these, you're not ready to source.

After the Intake: The Confirmation Email

Send this within 4 hours of the intake call:

"Based on our conversation, here's what I'm screening for:

First 30-day priority: [specific deliverable] **Top 3 must-have skills:** [not 15, just 3] **Success pattern:** [what good looks like] **Failure mode to avoid:** [what to filter out]

First submittals by [date]. Let me know if I've missed anything."

This does two things: 1. Confirms alignment before you waste time sourcing 2. Creates documentation when the hiring manager changes requirements later

When the Hiring Manager Can't Answer

Sometimes they don't know what they want. That's information too.

If they can't describe the first 30 days: "It sounds like the role is still being defined. Would it help to wait until there's more clarity before we start sourcing?"

If every requirement is must-have: "If I bring you a candidate who's strong in 3 of these but weaker in the others, would you interview them? That helps me understand true priorities."

If they describe perfection: "That profile would command $200K in this market. Is that within range, or should we discuss trade-offs?"

Your job isn't to be a yes-person. It's to surface reality so you don't waste weeks on an unfillable req.

For agency recruiters without direct client relationships: flag unclear requirements to your account manager with the data. Let them decide how to position the conversation with the client. Your job is surfacing the issue; the account relationship determines how it's communicated.

→ **Intake confirmation template:** See Appendix Section 4 (Client Templates).

The Unfillable Req Conversation

You've screened 40 candidates. Submitted 8. All rejected.

The pattern is clear: the requirements don't match the market. The salary is too low, the skill combination is too rare, or the expectations are unrealistic.

You have two options:

Option 1: Keep sourcing, hoping the perfect candidate appears. (They won't.)

Option 2: Have the difficult conversation with the client about why this req isn't working.

Most recruiters choose Option 1 because Option 2 feels risky. What if the client gets mad? What if they give the req to another agency?

Here's the reality: if you keep submitting candidates who get rejected, the client loses confidence in you anyway. The only difference is whether you lose credibility slowly (through failed submittals) or maintain it (by surfacing the problem and offering solutions).

When to Have the Conversation

Don't have this conversation on Day 3. You need data first.

The right timing: - You've sourced for at least 2 weeks - You've screened at least 15-20 candidates - You have a pattern of rejections or passes - You can articulate the market reality with specifics

Signs the req is unfillable: - Candidates consistently pass on comp - Every submittal is rejected for the same reason - The skill combination doesn't exist at this salary level - Market data shows the role is mispriced

The Conversation Framework

Step 1: Lead with data, not complaints

× "I'm having trouble finding candidates." √ "I've screened 25 candidates in the past two weeks. Here's what I'm seeing..."

Step 2: Present the pattern

"Of the 25 screened: - 8 passed on compensation (your budget is $120K; market rate is $145-160K) - 6 had the cloud experience but not the security background - 5 had the security background but not the cloud experience - 4 were interested but wanted remote; this role is hybrid - 2 were strong matches; both declined after seeing the tech stack"

Step 3: Offer options, not ultimatums

"Based on this data, I see three paths forward:

Option 1: Adjust compensation to $145K minimum. This opens up the candidate pool significantly.

Option 2: Adjust the seniority. At $120K, we can attract mid-level candidates who could grow into the role with mentorship.

Option 3: Pause the search until budget or requirements change. I don't want to keep submitting candidates who aren't the right fit."

Step 4: Let them choose

"Which direction makes the most sense for your team?"

What to Expect

Some clients will adjust. They'll appreciate the honesty and market insight.

Some clients will push back. "Other agencies are finding candidates."

(They're submitting unqualified people who will fail interviews too.)

Some clients will pause the req. That's not failure; it is saving everyone time.

Power Dynamics: When You're Junior

If you're early in your career, telling a senior hiring manager their req is unfillable feels risky. It is.

Tactics for navigating this:

Route it through your manager: "I've compiled this data on the search. Can we discuss how to present it to the client?"

Frame it as a question: "I want to make sure I'm targeting correctly. When candidates pass on $120K, is there flexibility, or should I focus on more junior profiles?"

Use market data as the authority: "The comp data shows this role at $145K-160K in this market. I wanted to flag that in case it helps explain the candidate responses."

The goal is surfacing reality without making it personal. Let the data do the work.

After the Conversation

Document what was decided. Send an email:

"Following up on our conversation: - We agreed to adjust compensation to $145K - Expectation: 2-3 qualified submittals within 10 days - I'll send a progress update by Wednesday"

This creates accountability and prevents "you never told me that" later.

The Solvable vs. Structural Problem

Solvable: Compensation is below market. Adjust it and candidates appear.

Structural: The role combines rare skills that don't exist together at any price point.

Solvable problems need one conversation. Structural problems may need a fundamental redesign: splitting the role, adjusting the team structure, or accepting that the unicorn doesn't exist.

Know which one you're dealing with before the conversation.

The Unfillable Req Readiness Checklist

Before having this conversation, verify:

√ You've screened for at least 2 weeks √ You have data on candidate pass/reject patterns √ You have market data from credible sources √ You've documented candidate pass reasons or rejection patterns √ You can present at least 2 options to move forward √ You've chosen the right timing (days 10-14) If all six are true, you're ready.

→ **No-Go decision framework:** See Appendix Section 9 (The No-Go Checklist).

Preventing Offer Dropouts

The client extends an offer. You call the candidate to share the good news. They sound excited.

Two days later: "I've decided to stay at my current company. They countered."

Or: "I accepted another offer. It came in yesterday and I had to decide fast."

Or: "I'm having second thoughts about the role."

Offer dropouts destroy placement rates. You've invested weeks of work in sourcing, screening, coordinating interviews, managing expectations. Then it falls apart at the finish line.

Most recruiters assume offer dropouts are bad luck or market competition. They're not. They're preventable.

This chapter shows you how to identify dropout risk early and intervene before the offer stage.

Why Offers Get Declined

There are three common reasons:

Reason 1: The candidate was never fully committed. They were exploring, comparing options, or using your process to get leverage for a counter-offer. You missed the signals during screening.

Reason 2: Something changed during the interview process. They saw a red flag (team dynamic, unclear role, disorganized process) that made them doubt the opportunity. You didn't catch it in time to address.

Reason 3: A competing offer came in and you didn't know they were interviewing elsewhere. They were not lying; they just did not volunteer the information. And you didn't ask.

All three are preventable with the right checkpoints.

The 3 Checkpoint Conversations

Most recruiters only talk to candidates at two points: after the screen and after the offer. That's not enough.

You need three checkpoints:

Checkpoint 1: After First Interview

Timing: Within 2 hours of their first interview

Purpose: Assess genuine interest and surface concerns early

The call: "How did the interview feel from your end? What stood out, positive or negative?"

What you're listening for:

Red flags: - "It was fine." (Neutral = not excited) - "I'm still exploring options." (Not committed) - "The role seems different than what we discussed." (Misalignment)

Green flags: - "I'm really excited about [specific aspect]." - "The conversation confirmed this is exactly what I'm looking for." - "I asked about [concern] and their answer made sense."

Why this checkpoint matters: If they're lukewarm after the first interview, they're unlikely to accept an offer later. You can either address their concerns now or move on to other candidates.

What to do if you see red flags: Ask directly: "On a scale of 1-10, how interested are you in moving forward?" If it's below 8, ask what would need to change to get to 10. If they can't articulate it, they're not serious.

Checkpoint 2: After Final Interview

Timing: Within 2 hours of their final interview

Purpose: Confirm commitment before the offer is extended and identify competing timelines

The call: "Sounds like the interviews went well. Before the client moves forward with an offer, I need to understand where you are:

1. If they extend an offer that meets your expectations, are you prepared to accept?
2. Are you interviewing anywhere else right now? If yes, what's the timeline?
3. Is there anything that would make you hesitate to accept?"

What you're listening for:

Red flags: - "I'd need to think about it." (Not ready to commit) - "I have a few other final rounds this week." (You're in a competition you didn't know about) - "I'm not sure about [specific aspect]." (Unresolved concern)

Green flags: - "If the offer is at $X and includes Y, I'm ready to accept." - "I'm interviewing one other place but this is my top choice." - "I'd want to review the full benefits package, but assuming it aligns, yes."

Why this checkpoint matters: You're forcing them to articulate commitment BEFORE the offer. If they're vague or hedging, you know dropout risk is high. You can either delay the offer until they're ready or inform the client of the risk.

What to do if you see red flags: Tell the client: "Candidate performed well, but I'm sensing hesitation. Recommend we give them 48 hours to finish other processes before extending an offer. If we extend now and they decline, we've burned our best candidate."

Checkpoint 3: After Offer Extended

Timing: Immediately after delivering the offer, then daily until acceptance

Purpose: Prevent counter-offers and competing offers from derailing acceptance

The initial call: "Congratulations! Now that you have the offer in hand, walk me through your next 48 hours. Who are you talking to? What are you thinking through?"

What you're listening for:

Red flags: - "I need to discuss with my current manager." (Counter-offer risk) - "I'm waiting to hear back from another company." (Competing offer risk) - "I need to think about whether this is the right move." (Cold feet)

Green flags: - "I'm discussing with my spouse tonight and will confirm tomorrow." - "I need to review the benefits details, but I'm planning to accept." - "Can you help me think through how to resign professionally?"

Why this checkpoint matters: The 24-72 hours after an offer is extended is when candidates talk to their current employer, receive counter-offers, and second-guess their decision. Daily contact keeps you in the loop and lets you intervene.

The Counter-Offer Conversation

Many candidates receive counter-offers. Your job is to counsel them BEFORE they talk to their current employer.

The pre-counter-offer script:

"Before you resign, your current employer may counter-offer. That's common, and it's worth thinking through in advance.

Some questions you might ask yourself:

- What were my original reasons for exploring this opportunity? Does the counter-offer address those underlying factors, or just the compensation piece?
- If my current employer could make these changes now, what prevented them before?
- How do I feel about the relationship after this process, and how might my employer view my tenure going forward?

There's no universal right answer. Some people accept counter-offers and thrive. Others find that the original issues resurface. The key is making the decision based on your original motivations, not just the numbers on the table.

Whatever you decide, I'll support it. This is your career."

Why this works: You're not pressuring them. You're giving them framing to think clearly.

Reading the Signals During the Process

Dropout risk exists on a spectrum. Your job is to assess where each candidate falls and adjust accordingly.

Low dropout risk: - Responds to messages within 2-4 hours - Asks detailed questions about the role, team, growth path - References specific conversations from interviews - Volunteers information about other processes - Talks about logistics (start date) like they're planning to accept

Medium dropout risk: - Responds within 24 hours but not faster Asks generic questions (benefits, PTO) - Gives positive but general feedback ("interview went well") - Mentions other interviews when asked but doesn't elaborate

High dropout risk: - Takes 48+ hours to respond to messages Doesn't ask questions about the role - Gives vague feedback ("still thinking about it") - Avoids discussing other interviews - Hesitates when you ask about commitment level

When you see medium or high dropout risk, increase checkpoint frequency and dig deeper on concerns.

What to Do When They Decline Anyway

Sometimes you do everything right and they still decline. That's not failure; it is market reality.

When it happens:

Ask for honest feedback: "I want to understand what happened so I can improve. What was the deciding factor?"

Document the pattern: If 3 candidates in a row decline for the same reason (comp, team concern, counter-offer), there's a systemic issue to address with the client.

Don't burn the relationship: "I'm disappointed it didn't work out, but I respect your decision. If things change in 6 months, I'd love to reconnect."

Many declined candidates come back within a year. Keep the door open.

The Dropout Prevention Checklist

For every candidate who reaches offer stage, verify:

✓ You've had all 3 checkpoint conversations ✓ You've asked about competing interviews and timelines ✓ You've counseled them on counter-offers before resignation ✓ You've assessed dropout risk level and adjusted accordingly ✓ You've maintained daily contact after offer extended

If all five are true, you've done your job.

→ **Checkpoint scripts and counter-offer language:** See Appendix Section 4 (Candidate Templates).

The System in Action

The previous chapters gave you the framework. This chapter shows you what it looks like when the framework meets reality.

These are composite scenarios drawn from real searches, with names changed, details altered, but the dynamics preserved. Each one demonstrates the system under pressure: the intake questions, the filtering decisions, the client conversations, and the judgment calls that determine whether a search succeeds or fails.

Read these as training simulations. Notice where the recruiter's internal reasoning diverges from what they say out loud. Notice when they push back and when they don't. Notice how they use data to make decisions rather than hope.

SCENARIO 1: The MarTech Architect

When the unicorn doesn't exist at your price point

The Setup

The req lands with urgency: MarTech Architect, hybrid in Dallas or New Jersey. CDP experience required. Real-time personalization. Tealium preferred, Adobe stack acceptable. Must understand GCP, Kafka, Pega integration. SFMC or Marketo. APIs and SQL. Oh, and the client wants AI/ML depth for personalization use cases.

Budget: competitive but not top-quartile. Timeline: yesterday.

This is either a $220K role or a fantasy. Let's find out which.

The Simulation

The intake call happens fast.

"Walk me through what this person does in their first 30 days."

Hiring manager: "They'll audit our current CDP implementation and design the integration architecture for our new personalization engine. We're migrating from batch to real-time."

Good. Specific deliverable. Not just "own the MarTech stack."

"What's the hardest part of this role that candidates underestimate?"

"The Tealium piece. Everyone claims Adobe. Almost no one has done Tealium at scale: event streams, consent management, audience segmentation. That's the gap."

There it is. Tealium expertise at this level is rare. Adobe profiles are everywhere. The Venn diagram overlap is tiny.

"If you had to cut the requirements in half, what survives?"

Pause. "Tealium. The CDP architecture experience. Everything else we can work around."

Now you know the filter. Tealium is the gate, not a nice-to-have.

Two weeks in. Sourcing has been aggressive: LinkedIn Recruiter, direct outreach, network mining. The pipeline tells a story:

- 12 candidates with strong Adobe backgrounds. Zero Tealium depth.

- 4 candidates with Tealium exposure. All junior implementations, not architecture-level.

- 2 candidates with the full stack. One wants full remote (non-starter for hybrid). One's rate is 40% above budget.

The market is speaking. This skill combination at this price point doesn't exist in sufficient quantity.

Screen notes from the strongest candidate:

"Tell me about your Tealium implementation. What was your specific responsibility?"

"We used Tealium for our tag management and event tracking. I worked with the team to configure the data layer."

"Worked with the team." "Configure." This is implementation, not architecture. They touched Tealium; they didn't design the system.

"What trade-offs did you navigate when choosing Tealium over a competing CDP?"

"That decision was made before I joined. I inherited the stack." *No architectural ownership. Pass.*

The client calls for a status update.

"I've screened 18 candidates in two weeks. Here's the pattern: Adobe expertise is abundant. Tealium architecture experience at the level you need is scarce. The two candidates who have it are either not open to hybrid or priced 35-40% above your range."

Silence.

"I see three options. One: we adjust compensation to attract the candidates who have this skill set. Two: we soften the Tealium requirement and hire someone who can learn it, accepting a longer ramp. Three: we pause until the market shifts or budget opens up."

The hiring manager sighs. "Let me take this back to leadership. The budget isn't mine to change."

Translation: this req is about to go on hold or stall.

The Decision Moment

A week later, the email arrives: "We're closing the MarTech Architect requisition. Budget reallocation. Thanks for your work on this."

Outcome: No-Go (External Factor)

The search didn't fail because of bad filtering. It failed because the requirement and budget were misaligned with market reality. The recruiter's job was to surface that reality early, which they did.

Post-Game Analysis

What worked: - **Intake question #4** ("cut requirements in half") revealed Tealium as the true gate. Without this, the recruiter would have wasted time on Adobe-only candidates. - **Data-driven client update** at day 14 forced a decision rather than letting the search drift for months. - **Three options, not complaints.** The recruiter framed the conversation around choices, not problems.

What to watch for: - **Niche platform skills** (Tealium, Pega, specific CDPs) are binary. Candidates either have depth or they do not; there is no "close enough." - **Hybrid requirements** in a remote-preference market cut your pool significantly. Always surface this trade-off at intake. - **When a req dies, redirect fast.** The recruiter had two strongish candidates who didn't fit this role. Within 24 hours, both were submitted to adjacent reqs at other clients.

SCENARIO 2: The Palantir Forward Deployed Engineer

When platform depth and clearance collide

The Setup

The ask: Forward Deployed Engineer with Palantir Foundry and AIP experience. Certifications preferred. Strong Python, PyTorch, SQL. The client is building an internal Palantir practice and wants someone who can land, deliver, and help seed the team. Billable by Q1.

Oh, and cleared candidates are a plus because there is government-adjacent work on the horizon.

Palantir talent is concentrated in a small ecosystem. Foundry depth isn't something you pick up from a YouTube tutorial. This search will be narrow but decisive: candidates either know the platform or they don't.

The Simulation

Two candidates surface quickly. Different profiles, same platform.

Candidate A: Forward Deployed background. Foundry certified. Recent AIP and Foundry deployments. Product and analytics experience. Client-facing. Currently available.

Candidate B: Data Engineer positioning. Foundry experience but no certification. Strong Python. Rate is lower. Less client-facing polish.

On paper, A is the clear front-runner. Let's validate.

Screen with Candidate A:

"Walk me through your most complex Foundry deployment. What was your specific responsibility?"

"I led the ontology design for a logistics client. Built the object types, defined the relationships, set up the pipeline configurations. When we integrated AIP, I owned the prompt engineering layer and the feedback loops."

"I led." "I built." "I owned." Ownership language. Good.

"What broke during implementation, and how did you diagnose it?"

"Pipeline latency. We were pulling from external transforms that weren't optimized. I used Grafana to trace the bottleneck, found we were over-fetching from the source system. Rebuilt the transform logic to pull incremental loads. Cut latency by 60%."

Specific diagnosis. Specific fix. Specific metric. This person did the work.

"What's your clearance status?"

"Not currently cleared, but I'm a U.S. citizen and clearable. Previous employer started the process but I left before it completed."

Noted. Not a dealbreaker for this req, but limits the government-adjacent upside.

Screen with Candidate B:

"Same question: walk me through your Foundry work."

"I was on a team that used Foundry for a supply chain project. I handled the data pipeline piece by writing transforms, managing the ingestion."

"On a team." "Handled." This is contributor-level, not architect-level.

"What trade-offs did you navigate in the ontology design?"

Pause. "The ontology was already designed when I joined. I worked within the existing structure."

No architectural ownership. This is a solid data engineer, but not an FDE.

"Tell me about your AIP exposure."

"Minimal. We were pre-AIP on that project."

The platform has evolved. This candidate is one generation behind.

Submission to the client partner:

"Submitting one candidate for the FDE role. [Candidate A] has the Foundry certification, recent AIP deployment experience, and led an ontology design from scratch. Client-facing background, available immediately, and strong technical depth. Rate is at the top of your range but justified by the profile.

[Candidate B] is a capable data engineer but doesn't have the FDE depth or AIP experience you need. Happy to keep them warm for future data engineering roles."

Client response: "Let's interview [Candidate A]. Can you coordinate?"

Interview coordination hits a snag. The client has an urgent conflict and needs to reschedule.

The recruiter's move: immediate outreach to the candidate.

"Interview needs to shift to 3 PM due to a client emergency. Are you flexible?"

Candidate: "Yes, I can make that work."

Don't let scheduling friction kill momentum. Fast response, fast resolution.

Post-interview, the candidate sends a thank-you note referencing specific topics: ontology best practices, debugging workflows, Grafana and Apollo usage. Signs off with genuine enthusiasm.

Candidates who reference specifics from the conversation are engaged. Generic thank-yous mean nothing.

The Decision Moment

Client feedback: "Strong interview. Moving forward. Let's discuss next steps."

Outcome: Go

Post-Game Analysis

What worked: - **10-minute screen questions** separated FDE-level ownership from contributor-level participation. Both candidates had "Palantir experience." Only one had depth. - **Platform-specific follow-ups** (ontology design, AIP integration, transform optimization) exposed whether candidates were current with the platform's evolution. - **Single strong submittal** rather than flooding the client with options. Quality over volume.

What to watch for: - **Niche platform searches** reward deep screening, not wide sourcing. Ten mediocre Palantir profiles are worth less than one validated FDE. - **Clearance is a timeline, not a binary.** "Clearable" candidates can work non-cleared projects while the process runs. Know your client's actual requirement vs. their wish list. - **Post-interview engagement signals matter.** Candidates who reference specifics are mentally committed. Those who go silent are hedging.

SCENARIO 3: The Internal Presales Hire

When the client is your own company

The Setup

This one's different. The req is internal: Presales Client Partner for the AI & Data practice. Reports to the practice leader. Base $180K plus $40K variable tied to delivery ownership, footprint expansion, and practice growth metrics.

Internal hiring looks like external hiring until it doesn't. The stakeholders are colleagues, not clients. The politics are closer. The comp structure has KPIs you actually have to understand.

The Simulation

Internal intake with the practice leader:

"What does success look like in the first 90 days?"

"They're landing at least two net-new client conversations per month. They're embedded in our delivery teams enough to spot expansion opportunities. They're not just selling; they are seeing where AI and data solutions solve problems the client hasn't articulated yet."

Presales that understands delivery. Not a pure sales hunter. This is a hybrid profile.

"What's the variable comp tied to?"

"Three buckets: Delivery Ownership means they stay close to projects they sell, not just toss over the wall. Footprint Expansion means growing existing accounts. Strengthening AI & Data means contributing to practice IP, thought leadership, internal enablement."

The KPIs are relationship-heavy. This person needs to be trusted by delivery, not just effective at closing.

"Who failed in this role before, and why?"

Pause. "We had someone who was a great closer but couldn't maintain credibility with the technical teams. Sold things we couldn't deliver. Burned trust internally. Lasted eight months."

There it is. The hidden filter isn't sales ability; it is technical credibility.

Candidate sourced internally: someone already in the network with presales and practice-building background.

Screen:

"Tell me about a deal you closed where you stayed involved post-sale. What was your role in delivery?"

"Healthcare AI project. I sold it, but I also sat in on the first three sprint reviews. When scope started creeping, I was the one who went back to the client to renegotiate rather than letting the delivery team absorb it. Protected the margin and the relationship."

Stayed involved. Protected the team. This is the anti-pattern to "sold and disappeared."

"How do you build credibility with technical teams who are skeptical of sales?"

"I don't pretend to be an engineer. But I learn enough to ask good questions and to know when a client ask is reasonable versus fantasy. And I never promise timelines without checking with delivery first. They've seen me say 'I need to confirm that' to a client in real-time.

That's where trust comes from."

Self-aware. Doesn't oversell. Checks with delivery. This is the profile.

Internal coordination: offer parameters confirmed with HR, leadership alignment verified, reporting structure finalized.

Then the wrinkle: device setup.

"Start date is contingent on laptop provisioning. Policy says no start until setup is complete."

The candidate is ready to go. IT needs five more business days.

Internal bureaucracy can kill momentum just like external bureaucracy. Don't promise a start date you can't control.

The recruiter's move: transparency with the candidate.

"Your offer is ready. Start date will be confirmed once IT completes device setup, expecting five business days. I'll update you daily until we have a locked date."

Candidate: "Understood. Appreciate the heads-up rather than a surprise delay."

No one likes delays. Everyone appreciates honesty about them.

The Decision Moment

Offer sent. Accepted same day.

Outcome: Go

Post-Game Analysis

What worked: - **Understanding the internal KPIs** before sourcing. Variable comp isn't just a number, it signals what the company actually values. - **Failure pattern question** revealed the hidden filter: technical credibility with delivery teams. A pure sales profile would have failed the same way the last hire did. - **Proactive communication on the device delay** maintained trust. Internal hires have fewer competing options, but they can still get cold feet if the process feels disorganized.

What's different about internal hiring: - **Stakeholders are permanent.** Burn trust with an internal hiring manager and you'll see them in meetings for years. External clients rotate; colleagues don't. - **Comp structures are more complex.** Variable tied to KPIs means you need to understand what those KPIs actually measure and whether the candidate can hit them. - **Operational gates are stickier.** Device provisioning, system access, internal training cannot be rushed the way external onboarding sometimes can.

SCENARIO 4: The Agentic AI Developer

When technical evaluation separates real from rehearsed

The Setup

Internal practice build: Agentic AI Developer. The team is standing up an AI Center of Excellence and needs developers who can actually build, not just talk about building. Hiring manager wants clarity on scope: Enterprise Architect or Solutions Architect? GenAI and TOGAF requirements are negotiable. Telecom domain experience is a plus.

Internal reqs come with internal politics. The scope isn't fixed because the stakeholders haven't fully agreed. Your job is to clarify before you source, not after you submit.

The Simulation

Intake with the hiring manager:

"Before I source, I need to understand: are we hiring an EA who sets strategy, or an SA who builds solutions?"

Pause. "Honestly, we need execution more than strategy right now. Someone who can design and code agentic workflows. The EA title was aspirational."

There it is. Title inflation in the job description. If you source EAs, you'll get strategists who don't code. The req needs an SA who builds.

"GenAI experience: is that a must-have or a plus?"

"Plus. We can train GenAI specifics. What we can't train is solid Python fundamentals and the ability to architect integrations."

The hidden filter isn't GenAI. It's core engineering depth. GenAI knowledge without engineering fundamentals is ineffective for agentic systems.

"TOGAF?"

"Don't care. That was HR's addition. Ignore it."

Half the job description is noise. Now you know what actually matters: Python depth, integration architecture, ability to execute. Telecom domain is a bonus, not a gate.

Two candidates reach final evaluation. Both claim agentic AI experience. Time to validate.

The interview includes a coding exercise. Two tasks:

Task 1: Write a function that returns elements at even indices from a list.

Task 2: Implement a custom sort without using built-in sort functions.

These aren't hard problems. They're filter problems. Senior developers solve them in minutes. Candidates who've been coaching their way through interviews stumble on the basics.

Candidate A:

Task 1 completed in 90 seconds. Clean code, handled edge cases.

Task 2: Implements bubble sort from memory. Not optimal, but correct. When asked about trade-offs: "Bubble sort is $O(n^2)$. For production, I'd use merge sort or Python's Timsort. But you asked for no built-ins, so I went with what I could write quickly and correctly."

Knows the trade-offs. Prioritized correctness over cleverness. Self-aware about the choice.

Follow-up: "Walk me through an agentic workflow you've built."

"Built a document processing agent for a telecom client. The agent parsed incoming contracts, extracted key terms using an LLM, validated against business rules, and routed exceptions to human review. The tricky part was the feedback loop: when humans corrected the agent's extraction, we logged those corrections and used them to improve the prompts."

Specific system. Specific challenge (feedback loop). Specific solution. This is real experience.

Candidate B:

Task 1: Takes four minutes. First attempt has an off-by-one error. Corrects after prompting.

Task 2: Starts confidently, then stalls. "I usually use sorted() for this." Produces incomplete code after eight minutes.

Four years of "senior" experience and can't implement a basic sort. The resume is inflated.

Follow-up: "Walk me through an agentic workflow you've built."

"I worked on a team that was exploring agentic AI for customer service. We were evaluating different LLM providers and building proof-of-concepts."

"Worked on a team." "Exploring." "Evaluating." "Proof-of-concepts." No production system. No ownership. This is exposure, not experience.

Feedback compiled and circulated internally:

"Candidate A: Strong recommendation. Solid Python fundamentals demonstrated in live coding. Articulated real agentic system with production complexity. Telecom domain experience is a bonus.

Candidate B: Do not proceed. Coding fundamentals below expectations for the level. Agentic AI exposure is conceptual, not hands-on. Resume overstates experience."

The Decision Moment

Candidate A receives offer. Device setup initiated. Start date set for after provisioning complete (internal policy: "start date is always after setup is completed").

Candidate B archived with notes. If they apply again in two years with more execution experience, reconsider.

Outcome: Go (Candidate A) / No-Go (Candidate B)

Post-Game Analysis

What worked: - **Scope clarification at intake** prevented sourcing the wrong profile. The JD said EA; the actual need was SA. Ten minutes of intake questions saved weeks of wrong-fit sourcing. **Live coding exposed the gap** that resumes hid. Four years of "senior experience" collapsed under a basic sort implementation. **Ownership questions separated real from conceptual.** "Worked on a team exploring" is not the same as "built and deployed."

What to watch for: - **Internal JDs often include aspirational requirements.** HR adds TOGAF. Hiring managers add GenAI. Ask what actually matters before sourcing. - **Coding exercises aren't about trick questions.** Simple tasks reveal whether fundamentals are solid. Candidates who struggle with basics will struggle with complex systems. - **Agentic AI is the new buzzword.** Everyone claims it. Few have built production systems. Ask for the feedback loop, the error handling, the edge cases. That is where real experience shows.

Applying the Scenarios

These four searches illustrate the same principles from different angles:

MarTech: When the market says no, surface it fast and give the client options. A req that stalls in week 3 is better than one that drifts for months.

Palantir: Platform depth beats keyword matches. Candidates with identical resume keywords had completely different ownership levels. The screen questions exposed the difference in 10 minutes.

Internal Presales: Internal hiring requires understanding the politics and KPIs that don't exist in external placement. The hidden filter was technical credibility, not sales numbers.

Agentic AI: Live coding separates real from rehearsed. Four years of "senior experience" can collapse under a basic sort implementation. Technical validation isn't optional for technical roles.

In all four cases, the outcome was determined by the quality of the intake questions, the precision of the screening, and the willingness to communicate transparently, even when the news wasn't good. That's the system in action.

Protecting Placement Integrity

Remote work changed the risk landscape. The same flexibility that expanded your talent pool also created new vulnerabilities: proxy interviews, AI-assisted misrepresentation, and undisclosed dual employment.

These aren't edge cases. They're patterns that experienced recruiters encounter regularly and patterns that damage client trust when they slip through screening.

Your job is verification, not accusation. The goal isn't to treat every candidate as suspect. It's to build screening practices that surface misrepresentation early, protecting your client, your credibility, and the legitimate candidates who deserve the role.

Proxy Interviews

The scenario: Someone other than the candidate attends the interview on their behalf. A technically strong friend, a paid stand-in, or a "interview service" that guarantees results. The real candidate shows up on Day 1 and cannot do the job.

Red flags during screening:

Camera "broken" for every interview. Voice or accent inconsistency across multiple calls. Knowledge gaps that don't match resume depth. Audio delays suggesting real-time coaching. Different energy or communication style between screens.

Prevention tactics:

Camera-on policy stated upfront. Make it non-negotiable. Multiple interviewers across different days; compare notes on voice, mannerisms, consistency. Live screen-share coding exercises are hard to coach someone through while they're sharing.

Resume-specific questions only the real candidate could answer: "Walk me through the architecture decision you mentioned at Company X."

Verification at offer stage:

Government ID verification matched to interview recordings. Day 1 video or in-person confirmation: same person, same voice, same knowledge level. Unannounced brief callback 24-48 hours after final interview: "Quick follow-up question from yesterday." The same person should answer the same way.

Contractual safeguard:

Include in offer letter: "Candidate confirms that all interviews were conducted by the candidate personally. Identity misrepresentation is grounds for immediate termination and may result in forfeiture of compensation."

AI-Assisted Misrepresentation

The scenario: The candidate uses real-time AI tools (ChatGPT, interview assistants, or similar) to answer technical questions during live interviews, presenting AI-generated responses as personal knowledge and experience.

The issue isn't the tool. Candidates using AI to prepare for interviews is legitimate. The misrepresentation is claiming AI-generated answers as lived experience. Your job is designing interviews that reveal the difference.

Red flags during screening:

Unnaturally comprehensive, textbook-perfect answers. Real engineers hedge and qualify. Pauses that feel like reading, not recalling. Eye movement patterns suggesting reading from a second screen. Strong conceptual answers that collapse under follow-up depth. Responses that sound like documentation rather than experience. Keyboard clicking during audio-only portions.

Prevention tactics:

Conversational depth that AI can't maintain: Start broad ("Tell me about X"), go narrow ("Why did you choose Y?"), then to edge cases ("What broke and how did you fix it?"). AI generates plausible first answers but struggles with the third level of depth.

Experience-specific questions tied to their resume: "You mentioned building the payment system at Company X. Walk me through a specific decision you made that you later regretted." AI can't fabricate regret about projects it didn't build.

Ask for failures, mistakes, and opinions. AI hedges; engineers commit. "What's the most overrated technology in your stack?" "What would you refuse to use again?" Real practitioners have opinions. AI gives balanced non-answers.

Live screen-share with audio. It's difficult to read prompts while sharing your screen and narrating your thought process. Watch for minimized windows, unusual tab-switching, or "let me think" pauses that coincide with typing sounds.

The three-level depth rule:

Level 1: "How does Kubernetes handle pod scheduling?" (AI answers well) Level 2: "You mentioned using Kubernetes at your last role. What scheduling constraints did you configure and why?" (AI struggles without real context) Level 3: "What went wrong with that configuration and how did you debug it?" (AI can't fabricate authentic debugging stories)

If a candidate aces Level 1 but stumbles on Levels 2 and 3, the knowledge isn't theirs.

Undisclosed Dual Employment

The scenario: The candidate holds multiple full-time positions simultaneously without disclosure. Your client pays for dedicated focus; they're getting split attention. Deliverables slip. Availability becomes erratic. By the time the pattern is clear, the placement is already damaged.

Red flags during screening:

Immediate availability despite claiming current full-time employment. No notice period needed? Vague about current role responsibilities. Reluctance to discuss transition timeline. Already "flexible" on working hours before the job starts. Pushback on standard background verification.

Red flags during engagement:

Consistent calendar conflicts during core hours. Background noise or notifications suggesting concurrent meetings. Productivity gaps not explained by workload. Repeated declined camera requests. Working odd hours, unresponsive during standard business hours. Multiple Slack or Teams instances visible during screen shares.

Prevention at screening:

Ask directly: "Do you currently have any other employment, consulting, or contractor commitments?" Document the answer in your notes. Verify current employment status through background check. Confirm notice period. Immediate availability for someone "currently employed full-time" warrants follow-up questions.

Contractual safeguards:

Exclusive employment clause: "Employee confirms this position is their sole full-time employment during the engagement period."

Disclosure requirement: "Any outside employment, consulting, or contractor work must be disclosed in writing and approved in advance."

Right to verify: "Employer reserves the right to verify exclusive employment status during the engagement."

Resolution if discovered:

Document the pattern first: dates, incidents, evidence. Direct conversation before escalation: "We've noticed availability gaps and want to understand what's happening." If confirmed: client notification per contract terms, termination with documentation. If suspected but unconfirmed: increase check-ins, tighten deliverable timelines, monitor closely.

The Verification Mindset

Trust but verify isn't cynicism. It's professionalism.

Most candidates are exactly who they claim to be. Your verification practices aren't designed to catch the majority; they are designed to catch the minority before they damage your client relationship and your reputation.

The recruiters who build long careers treat verification as routine, not interrogation. Camera-on policies. Depth-based questioning. Direct questions about employment status. These become standard practice, applied consistently to everyone, not selectively based on suspicion.

Your job is protecting three parties: the client who's investing in talent, your own credibility as a trusted partner, and the legitimate candidates who deserve roles they've genuinely earned. Verification serves all three.

Appendix: The Complete Toolkit

You've read the chapters. You understand the mindset, the skills, the principles.

Now you need the scripts, the templates, and the checklists you can use Monday morning.

This appendix is your reference guide. Bookmark it. Use it when you need a specific tool.

SECTION 1: RESUME FILTERING CHECKLISTS

90-Second Resume Scan Checklist

SCAN 1: Job Titles (5 seconds) - [ ] Does progression make sense? [ ] Are titles inflated? - [ ] Are there unexplained gaps?

SCAN 2: Current Role (30 seconds) - [ ] Are outcomes specific? - [ ] Is ownership clear? - [ ] Does complexity match the title?

SCAN 3: Skills Section (10 seconds) - [ ] Reasonable number of technologies? - [ ] Depth in one stack vs superficial breadth?

SCAN 4: Red Flags (30 seconds) - [ ] Buzzword overload without context? - [ ] Vague ownership language? - [ ] Job-hopping without pattern?

SCAN 5: Decision (15 seconds) - [ ] Screen / Maybe / Pass

SECTION 2: SCREENING QUESTIONS (Core 5 for Any Role)

Q1: Ownership "Walk me through the most complex project you've worked on. What was YOUR specific responsibility?"

Q2:   Trade-offs "Why did you choose [Technology A] over [Technology B]?"

Q3: Failure Handling "Tell me about something that broke in production. What was the impact and how did you diagnose it?"

Q4: Evolution "If you rebuilt that system today, what would you do differently?"

Q5: Decision-Making "What's the hardest technical decision you've made in the last year?"

SECTION 3: PATTERN RECOGNITION GUIDE

Dimension	Deep Experience	Surface Experience
Ownership	"I designed," "I built"	"We implemented," "Involved in"
Specificity	Numbers, constraints, timelines	Abstract descriptions
Trade-offs	Acknowledges downsides	Decision seems obvious
Failure	Takes ownership, learned	Blames others
Evolution	"I'd do X differently because..."	"I'd do the same"
Business context	Connects tech to business	Stays purely technical

SECTION 4: COMMUNICATION TEMPLATES

Candidate Templates

Initial Acknowledgment: "Thanks for your interest in [role] at [company]. I'll review your background by [time] and reach out if there's a strong match."

Post-Screen Pass: "Thanks for speaking with me about [role]. After reviewing your background against the requirements, I don't think this is the right match. I'll keep you in mind for future roles."

Post-Screen Move Forward: "Great conversation today. I think you'd be a strong fit for [role]. Next step: I'm submitting your profile to the client by [date]."

Interview Confirmation: "Interview confirmed for [Day, Date] at [Time] with [Name/Title]. Key things to highlight: [2-3 points]."

Interview Reminder: "Reminder: Interview tomorrow at [Time] with [Name]. All set?"

Offer Delivery: "Great news - [Company] would like to extend an offer! Details: [Title, Salary, Start Date]. Let's schedule a call to walk through everything."

Counter-Offer Counsel: "Before you resign, your current employer may counter-offer. That's common. Some questions worth thinking through: What were your original reasons for exploring this opportunity? Does the counter-offer address those underlying factors? Whatever you decide, I'll support it. This is your career."

Client Templates

Intake Confirmation: "Based on our conversation, here's what I'm screening for: [First 30-day priority, Top 3 skills, Success pattern, Failure mode]. First submittals by [date]."

Candidate Submittal: "Submitting [#] candidates for [role]: [Name -

Current role - Key strengths - Why they're a match - Availability]."

Follow-Up (No Response): "Checking in - did the profiles for [role] come through? Happy to provide additional context."

Weekly Update (No Submittals): "Update on [role]: Screened [X] candidates. [Y] had skills but lacked [requirement]. [Z] passed on comp. Adjusting strategy to [approach]. Expect [X] submittals by [date]."

Unfillable Req (Solvable): "Update on [role]: Screened [X] candidates. Market reality: salary of $[X] is below market([Y]-[Z]). [X] of [Y] passed on comp. Options: 1) Adjust comp to [range], 2) Adjust seniority to [level], 3) Pause until budget adjusts. Which direction?"

Unfillable Req (Structural): "Candid conversation: Screened [X], submitted [Y]. All rejected for [reason]. This suggests misalignment between screening and interview expectations. Recommend calibration call to align on what [requirement] means behaviorally."

SECTION 5: 4-WEEK IMPLEMENTATION GUIDE

Week 1: Filter Calibration

Goal: Master 90-second resume scan and 10-minute screen

Daily: - Review 20 resumes using checklist - Screen 3 candidates using 5 core questions - Document patterns in strong vs weak answers

Checkpoint: Can you scan in 90 seconds and identify vague ownership in 2 minutes?

Week 2: Intake Discipline

Goal: Extract real requirements in 15 minutes

Daily: - Use 5 intake questions on every new req - Send confirmation email within 4 hours - Document hiring manager response patterns

Checkpoint: Are you getting clearer requirements? Are submittals passing at higher rates?

Week 3: Communication Speed

Goal: Implement 24-hour trust protocol

Daily: - Respond to applications within 4 hours (templates) - Send post-screen feedback within 24 hours - Send weekly client updates even if "no news"

Checkpoint: Are candidates responding faster? Are clients giving faster feedback?

Week 4: Full System Integration

Goal: Run complete process from intake to offer

Daily: - Apply all filters: intake, resume scan, screen, communication - Track: screens, submittals, pass-through rate

Checkpoint: Is time-to-fill decreasing? Is interview-to-placement improving? Working fewer hours for same results?

SECTION 6: QUICK REFERENCE CARDS

Daily Activity Checklist

Morning (30 min): - [ ] Respond to overnight applications - [ ] Review scheduled interviews - [ ] Send interview reminders

Mid-Day (2 hours): - [ ] Run 3-5 screens - [ ] Review new resumes (40 per hour)

Afternoon (1 hour): - [ ] Send post-screen feedback - [ ] Follow up on submitted candidates - [ ] Send client updates

End of Day (30 min): - [ ] Document screen notes - [ ] Plan tomorrow's priorities - [ ] Clear inbox

Weekly Planning Checklist

Monday: - [ ] Review all active reqs - [ ] Prioritize by urgency and fillability - [ ] Plan sourcing for top 3 reqs

Wednesday: - [ ] Send weekly updates to all clients - [ ] Review submittal pipeline - [ ] Follow up on pending interviews

Friday: - [ ] Analyze week's metrics - [ ] Identify what worked/didn't - [ ] Plan adjustments for next week

SECTION 7: TROUBLESHOOTING GUIDE

"Candidates keep failing interviews": Your screen questions aren't predictive. Compare notes to interview feedback. Ask client what specifically made candidate fail. Adjust questions.

"Client keeps rejecting for vague reasons": Intake wasn't thorough. Schedule calibration call using 5 intake questions. Get specific examples of "good" vs "not good." (Agency recruiters: work with your account manager to facilitate this call, or provide the questions for them to ask.)

"Candidates keep ghosting after offer": You're not running 3 checkpoint conversations. After first interview: assess interest. After final: confirm competing processes. After offer: daily contact.

"Working 60 hours/week, still missing quota": Optimizing for activity instead of results. Track time per req for one week. Identify unfillable reqs and cut them. Focus best hours on fillable reqs only.

"Hiring manager keeps changing requirements": They don't know what they want. Send confirmation email after every conversation. When they change again, reference email: "This is 3rd shift - can we lock criteria before continuing?"

FINAL NOTE: HOW TO USE THIS TOOLKIT

Don't memorize everything. Use this appendix as reference:

SECTION 8: QUICK DIAGNOSTIC QUESTIONS

When you need to assess depth in 60 seconds:

For any technical role: "What's the hardest technical problem you solved in the last six months, and why was it hard?"

Listen for: Specific constraints, trade-off reasoning, ownership language

For platform specialists: "If I put you in front of (Foundry/Tealium/etc.) tomorrow with a blank slate, what's the first thing you'd build to prove you know the platform?"

Listen for: Concrete starting point, architecture thinking, confidence vs. hesitation

For architects: "Tell me about a design decision you made that someone disagreed with. How did you handle it?"

Listen for: Reasoning under pressure, stakeholder management, outcome

For candidates claiming AI/ML experience: "Walk me through how you'd evaluate whether an AI solution is actually working in production."

Listen for: Metrics beyond accuracy, monitoring approach, feedback loops

For roles requiring active security clearance: "When was your last periodic reinvestigation, and what level is your current clearance?"

Listen for: Specific dates and levels vs. vague "I have a clearance"

- **Screening candidate?** See Section 2 for questions
- **Running intake?** See Section 4 for templates

- **Need to send an update?** See Section 4 for templates
- **Starting implementation?** See Section 5 for 4-week guide
- **Analyzing conversation?** See Section 3 for patterns

SECTION 9: THE NO-GO CHECKLIST

When to Stop a Search or Initiate the Unfillable Req Conversation

Not every req is fillable. Recognizing this early protects your time, your credibility, and your client relationships. Use this checklist after 10-14 days of active sourcing.

Structural Red Flags:

- Market data confirms role is mispriced by 20%+ and client won't adjust
- Required skill combination doesn't exist at this salary level
- (unicorn req)
- Candidate pipeline exhausted after 3 weeks of quality sourcing
- Clearance or eligibility requirements eliminate 90%+ of qualified pool

Process Red Flags:

- 3+ submittals rejected for reasons not discussed in intake
- Client requirements have changed 3+ times without stabilization
- Hiring manager can't articulate what success looks like (even after intake questions)
- You're spending 15+ hours/week on a single req with no movement

Decision Protocol:

If 2+ boxes are checked in either category:

1. Compile your data (screens completed, rejection patterns, market rates)

2. Prepare 2-3 options (adjust comp, adjust requirements, pause search)

3. Schedule the Unfillable Req Conversation (Chapter 6)

Remember: Surfacing an unfillable req early protects your credibility. Grinding on it for months destroys it.

Bookmark this appendix. Return when you need a specific tool. Build muscle memory over time.

The goal isn't perfection. The goal is progress. One better screen at a time. One clearer intake at a time. One faster response at a time.

That's how you move from Level 2 to Level 3.

SECTION 10: AI PROMPT TEMPLATES

Important: AI models evolve. These prompts work as of this writing, but you may need to adjust them as tools change. The principles matter more than the exact wording. Use these as starting points, not scripts.

1. Resume Summary for Quick Review

Use when: You have 50+ resumes and need to triage quickly.

Prompt: "Summarize this resume in 3 bullets: (1) Current role and years of experience, (2) Top 3 technical skills with depth indicators, (3) Any red flags (gaps, job-hopping, vague ownership language). Be direct."

2. Boolean String Generator

Use when: Starting a new search and need multiple sourcing angles.

Prompt: "Generate 5 Boolean search strings for LinkedIn Recruiter to find [role]. Include variations for: (1) exact title matches, (2) skill-based searches, (3) company-based searches targeting competitors, (4) certification-based searches, (5) tool/platform-specific searches. Format for copy-paste."

3. Outreach Message Drafting

Use when: Personalizing outreach at scale.

Prompt: "Draft a 75-word LinkedIn InMail for [candidate name] based on their profile. Reference one specific project or skill from their experience. Position: [role] at [company]. Tone: professional but conversational. End with a soft ask, not a hard sell. Do not use exclamation points or phrases like 'exciting opportunity.' "

Always edit before sending. AI drafts are starting points. Your voice and judgment finalize them.

4. Technical Concept Primer

Use when: You encounter unfamiliar technology in a req and need to screen intelligently.

Prompt: "Explain [technology/framework] in 2 paragraphs for a non-technical recruiter. Include: (1) What problem it solves, (2) What roles typically use it, (3) What a senior practitioner would know that a junior wouldn't, (4) One screening question I could ask to test depth."

5. Salary Benchmarking Query

Use when: You need market rate context for client conversations.

Prompt: "What is the current market salary range for a [role] with [X] years of experience in [location]? Provide: (1) 25th/50th/75th percentile ranges, (2) Key factors that move candidates to the higher end, (3) How remote vs. hybrid vs. onsite affects comp. Note your data sources and any limitations."

Verify against current job postings. AI comp data can lag. Cross-reference with live market signals.

6. Interview Prep Research

Use when: Prepping a candidate for a client interview.

Prompt: "Research [company] for interview prep. Provide: (1) Recent news or announcements (last 6 months), (2) Tech stack if publicly known, (3) Company culture signals from Glassdoor/LinkedIn, (4) Three intelligent questions the candidate could ask the hiring manager."

7. Job Description Analysis

Use when: Preparing for an intake call or identifying potential issues.

Prompt: "Analyze this job description: [paste JD]. Identify: (1) Must-have vs. nice-to-have skills (flag if unclear), (2) Potential unicorn requirements (rare skill combinations), (3) Red flags (unrealistic expectations, unclear scope), (4) Three clarifying questions I should ask the hiring manager."

8. Rejection Email Softening

Use when: You need to deliver bad news while preserving the relationship.

Prompt: "Rewrite this rejection message to be warmer while remaining honest: [paste draft]. Keep it under 100 words. Don't use clichés like 'we'll keep your resume on file.' End with a genuine door-opener for future contact."

9. Counter-Offer Risk Assessment

Use when: Candidate is in offer stage and you need to prep for retention pressure.

Prompt: "Based on what I know about this candidate ([current company], [tenure], [reason for leaving], [new offer details]), what counter-offer scenarios should I prepare them for? What questions should I ask to assess their true commitment level?"

10. Client Update Drafting

Use when: You need to send a progress update with no good news.

Prompt: "Draft a client update email for [role]. Situation: [X candidates screened], [Y rejected for specific reasons], [Z passed on comp]. Tone: transparent, professional, solution-oriented. Include: what I've done, what I've learned, what I'm adjusting. Under 150 words."

The Rule for All AI Prompts

AI handles the draft. You handle the judgment. Never send AI output without review. Never let AI make decisions that require context it does not have, such as candidate motivation, client politics, relationship history. The time you save on drafting should be reinvested in thinking, not skipped entirely.

Practice Mode: AI as Your Coaching Partner

Reading scripts is not the same as using them under pressure. AI can help you practice difficult conversations before they happen: privately, repeatedly, without judgment. Use these prompts to create roleplay scenarios. Make mistakes here, not with real candidates and clients.

Practice 1: The Evasive Candidate

Prompt: "You are a software engineer with 5 years of experience. You worked on a team that built a payments platform, but you were a contributor, not a decision-maker. When I ask ownership questions, deflect to team accomplishments. Use phrases like 'we built' and 'the team decided.' Make me work to uncover your actual individual contribution. Stay in character."

What you're practicing: The Ownership Test. Spotting vague language.

Follow-up questions that pin down individual contribution.

Practice 2: The Resistant Hiring Manager

Prompt: "You are a hiring manager who wants a Senior DevOps Engineer with Kubernetes, Terraform, AWS, Azure, GCP, security certifications, and 7+ years of experience. Budget is $130K. When I push back on requirements or compensation, resist. Say things like 'other agencies are finding candidates' and 'we need someone who can hit the ground running.' Make me practice the unfillable req conversation."

What you're practicing: Surfacing market reality. Presenting options without ultimatums. Holding your ground professionally.

Practice 3: The Counter-Offer Candidate

Prompt: "You are a candidate who just accepted my client's offer. Now your current employer has counter-offered: $15K raise, promotion to Senior, and a promise to fix the issues you complained about. You're wavering. When I counsel you, push back with 'but they're finally recognizing my value' and 'maybe I should give them another chance.' Make me practice counter-offer counseling."

What you're practicing: Asking the right questions. Helping them think clearly without pressuring. Supporting their decision either way.

Practice 4: The Vague Intake Call

Prompt: "You are a hiring manager who doesn't really know what you want. When I ask about first-30-day priorities, be vague: 'help the team' and 'own the backend.' When I ask what skills matter most, say 'all of them.' When I ask about past failures, say 'we just need someone good.' Make me work to extract real requirements."

What you're practicing: The 5 intake questions. Pushing past vague answers. Knowing when to pause a search until requirements clarify.

Practice 5: The Disappointed Rejection

Prompt: "You are a candidate I'm about to reject after final interviews. You really wanted this job. When I deliver the news, push back: ask why, express frustration, say 'I thought I nailed it.' Make me practice delivering rejection with honesty and empathy while preserving the relationship."

What you're practicing: Delivering bad news directly. Providing useful feedback without overpromising. Keeping the door open for future opportunities.

After Each Practice Session

Ask the AI: "Break character. How did I do? What could I have asked differently? Where did I miss an opportunity to dig deeper?"

This feedback loop is where the real learning happens. Practice, get feedback, adjust, repeat. The conversations that used to make you nervous will become routine.

A Note on Sources and Methodology

The frameworks in this book emerged from two decades of practice: thousands of placements, hundreds of team training sessions, and countless conversations with recruiters, clients, and candidates across multiple market cycles.

Where our thinking builds on established principles (behavioral interviewing methodology, trust-building communication research, competency-based assessment, structured screening techniques), we have adapted those foundations to the specific realities of IT agency recruiting. We're practitioners, not academics, and this book reflects that orientation: tested in the field, refined through repetition, presented without footnotes.

For readers interested in the research foundations underlying concepts like behavioral interviewing, structured assessment, or trust dynamics in professional relationships, those bodies of work are well-documented in the organizational psychology and HR literature. We encourage exploration, and we are confident that what you find will reinforce, rather than contradict, the practical system presented here.

About the Authors

Sejal Khimani and **Rahul Kale** are Senior Directors leading talent acquisition and technical recruiting, with a combined 30+ years of experience placing IT professionals.

Their partnership began at a Dallas-based IT consulting firm, where they spent years working side by side: screening candidates, navigating client relationships, and learning what separates recruiters who stay busy from those who consistently place. Both had started their careers as engineers with Master's degrees in Electrical Engineering. Neither planned to become a recruiter. Circumstance and opportunity led them here.

After years on separate paths (building teams, mentoring recruiters, progressing from individual contributor to management to senior leadership), they found themselves reunited, continuing to lead talent acquisition and resource development.

This book captures the system they built together: a framework refined across two decades, multiple market cycles, and thousands of placements. It's what they teach the recruiters they manage today: the same methods that have kept them building careers in an industry where most people burn out within the first few years.